Robotics Engineers

Kelly Daniels

Series Editor **Casey Malarcher**

Level 1 - ❶

Robotics Engineers

Kelly Daniels

© 2018 Seed Learning, Inc.

Series Editor: Casey Malarcher
Acquisitions Editor: Anne Taylor
Copy Editor: Liana Robinson
Cover/Interior Design: Highline Studio

ISBN: 978-1-943980-33-8

10 9 8 7 6 5 4 3 2
22 21 20 19

Photo Credits

All photos are © Shutterstock, Inc.

Contents

The Job of Robotics Engineers

Who makes robots?

Robotics engineers make them!

What kind of people make good robotics engineers?

Engineers working on a robot

Robotics engineers need to be creative.

Building a robot is not easy.

New problems come up all the time.

Robotics engineers also need to be patient.

It takes a long time to build a good robot.

Some robots take years and years to make!

Thinking about how to solve problems

Robotics engineers make robots for all different kinds of jobs.

They make robots to do hard work.

Strong robots can lift and move heavy things easily.

◀ A robot in a car factory

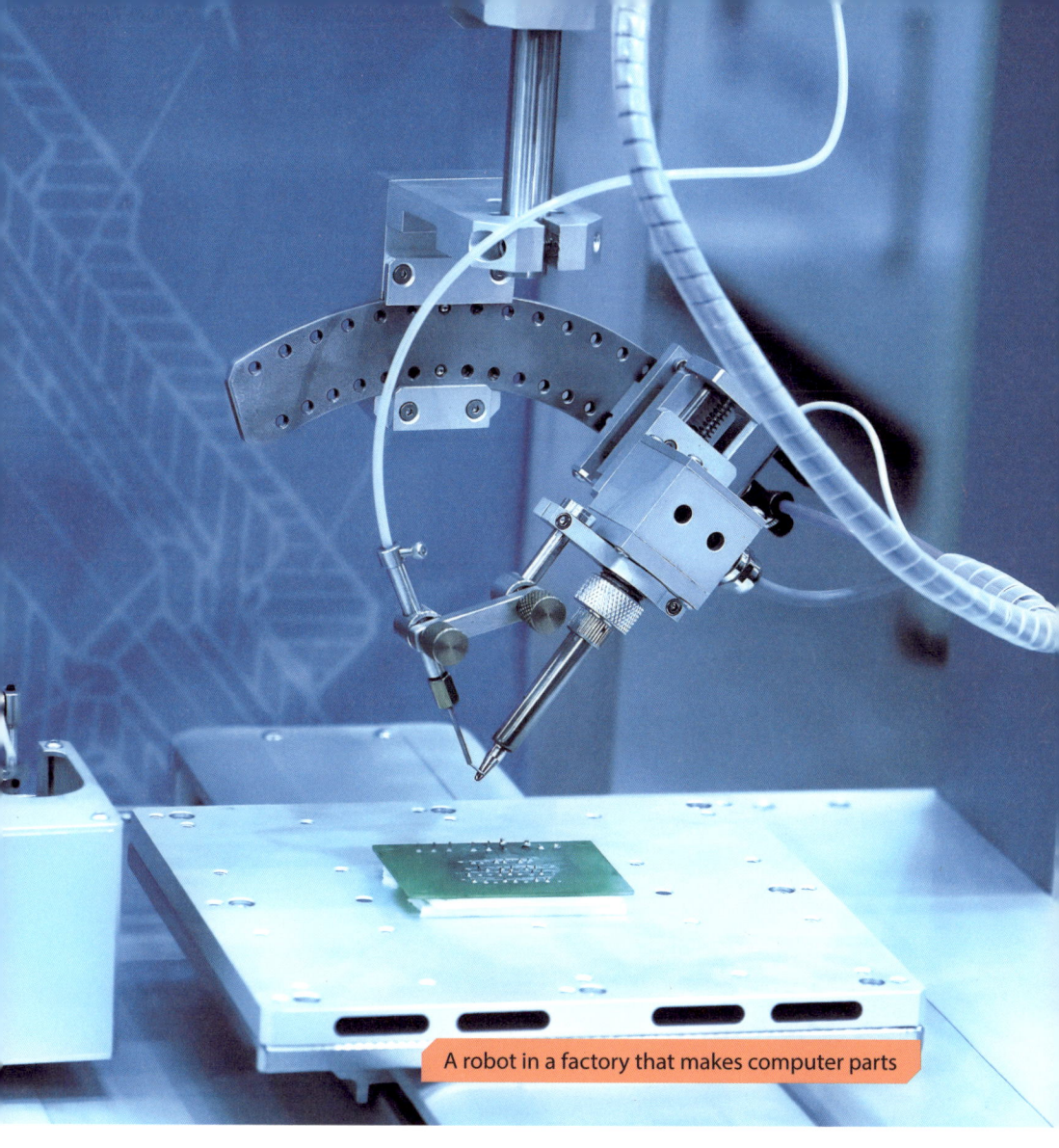

A robot in a factory that makes computer parts

Robotics engineers make robots to build things.

Some things have very small parts.

Robots can put the pieces together very carefully.

Robots make fewer mistakes than people.

Robotics engineers make robots to help people.
Some robots work in hospitals.
Doctors use them to help sick people.
These robots must be carefully made so they are clean and safe.

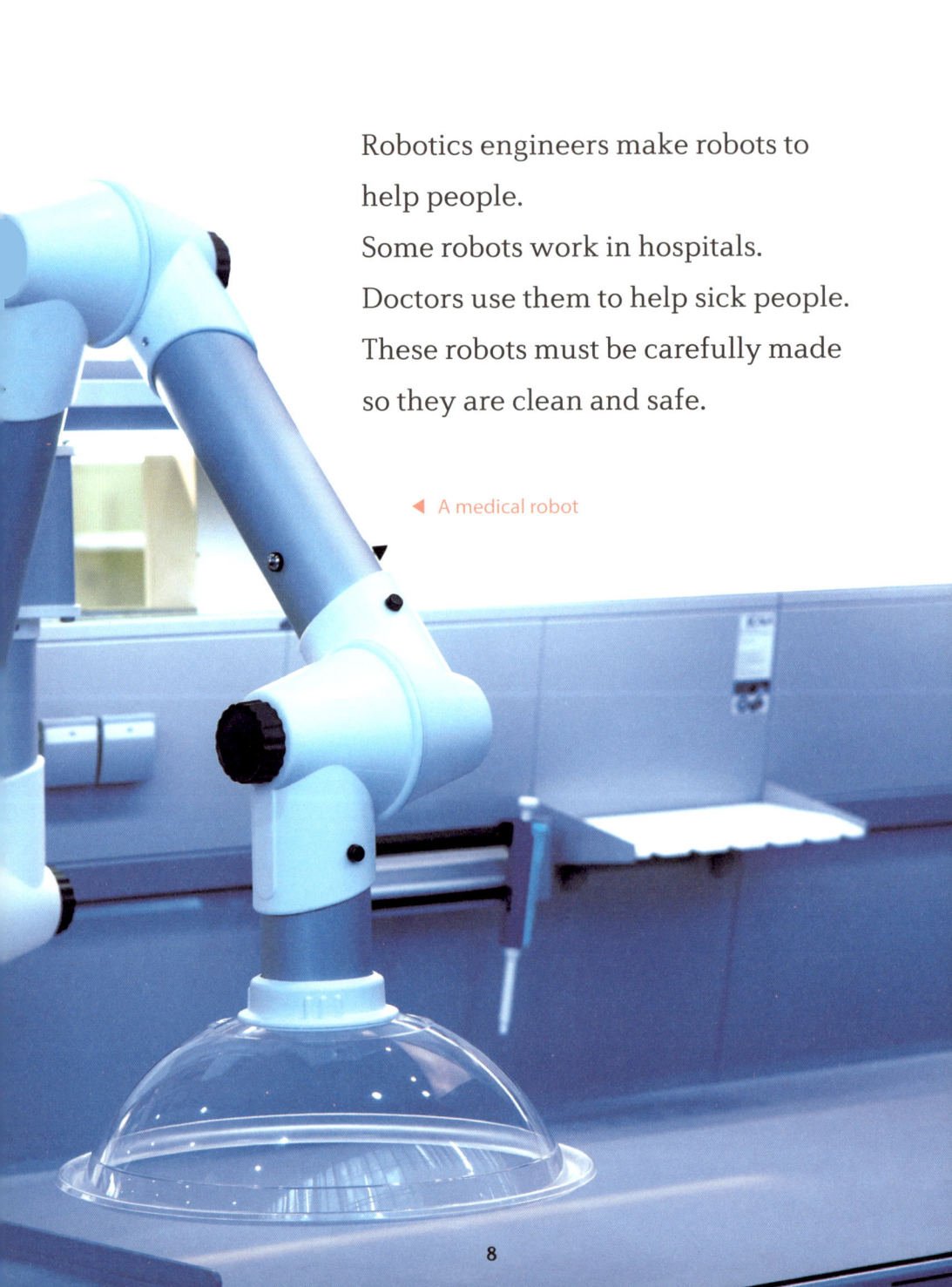

◀ A medical robot

Robotics engineers make robots to play with people.

Sometimes these robots play very well.

Today, computers can easily win at chess against people.

So computers play against each
other to see which is the best
in the world!

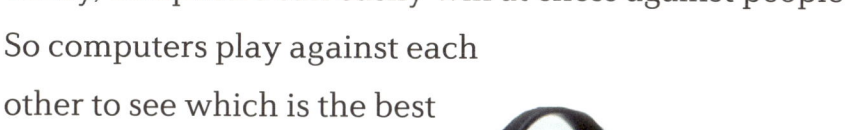

Playing chess ▶
against a robot

How to Be a Robotics Engineer

What do robotics engineers study?

In university, robotics engineers can study many things. But even before that, some students study robots in special programs.

◀ Learning skills in school to become a robotics engineer

A team working on a robot for a contest

There are contests for making robots.

Students can win money for school in these contests.

Robotics summer camp for kids

There are also robotics summer camps.

At these camps, kids learn about robots.

They also practice making robots of their own.

At universities, there are many things students can study.
They can study computers.
They can study science or math.
They can study engineering.
Any degree in one of these fields is useful for robotics
engineers.

University students in class

A four-year degree is enough to start work as a robotics engineer.

But with more study, you can start with a better job.

Some robotics engineers have masters degrees.

Other robotics engineers have Ph.D.s.

◄ Different university degrees that students study for

Working with a more
experienced engineer
in a company ▶

After university, a robotics engineer will usually begin
as an assistant engineer in a company.
An older, more experienced engineer will work with
him or her.

Robotics engineers need to learn how to work on teams. The people on the team listen to each other.

A team of robotics engineers working together

On a big team, some people work on only one part of the robot.

Other parts of the team work on different parts.

In the end, all parts of the team put their ideas together to make one robot.

A robot with parts made by different teams

Robots at Work

Many robotics engineers work in companies. The companies make things like cars. In fact, most robots today are found in the factories of car companies.

An engineer controlling ▶
a robot in a car factory

18

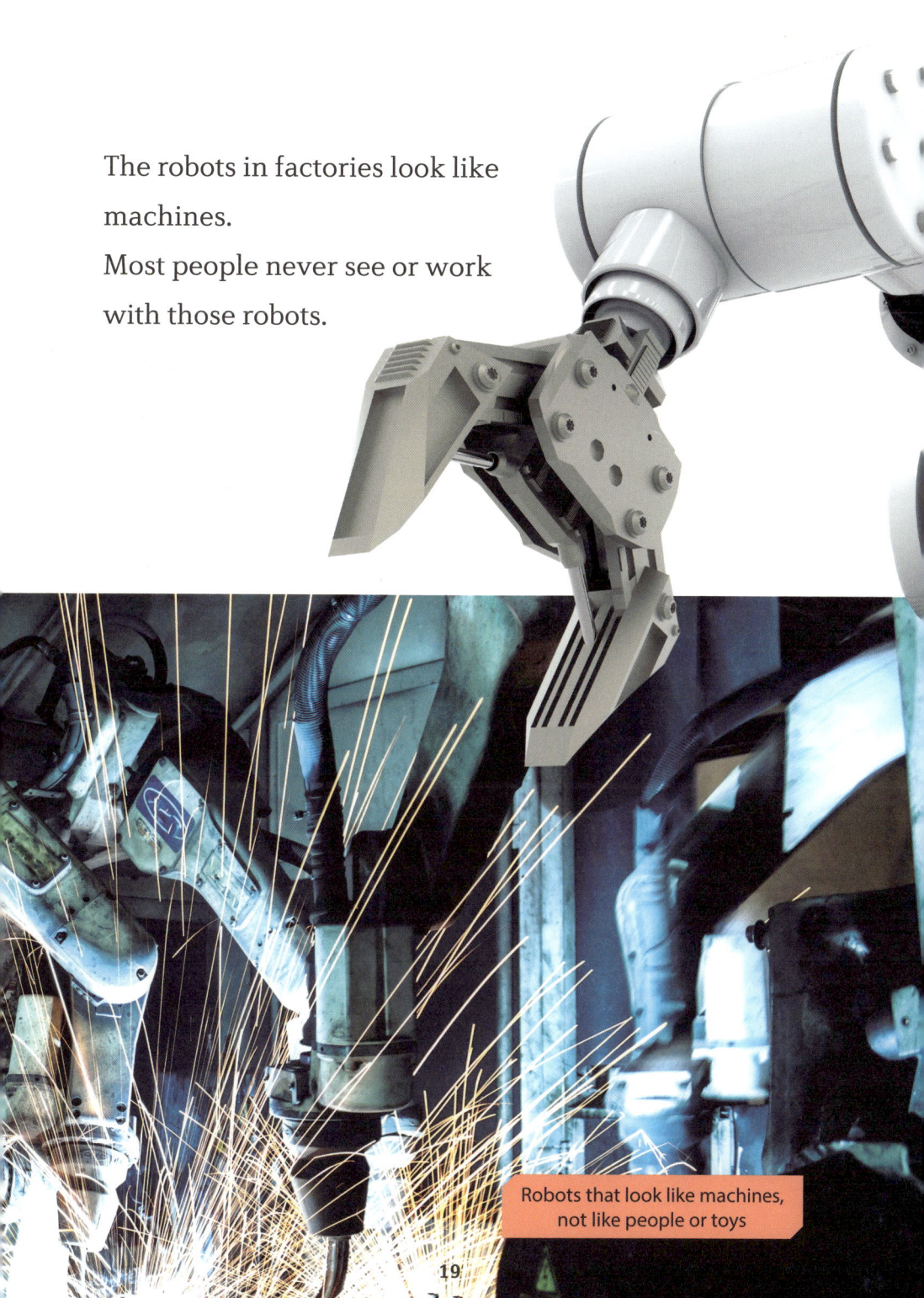

The robots in factories look like machines.
Most people never see or work with those robots.

Robots that look like machines, not like people or toys

19

But in other places, people may see robots every day.

Those robots need to look nice.

They need to look easy to use.

Then people will like to have them around.

A man controlling a ▶
cleaning robot in his home

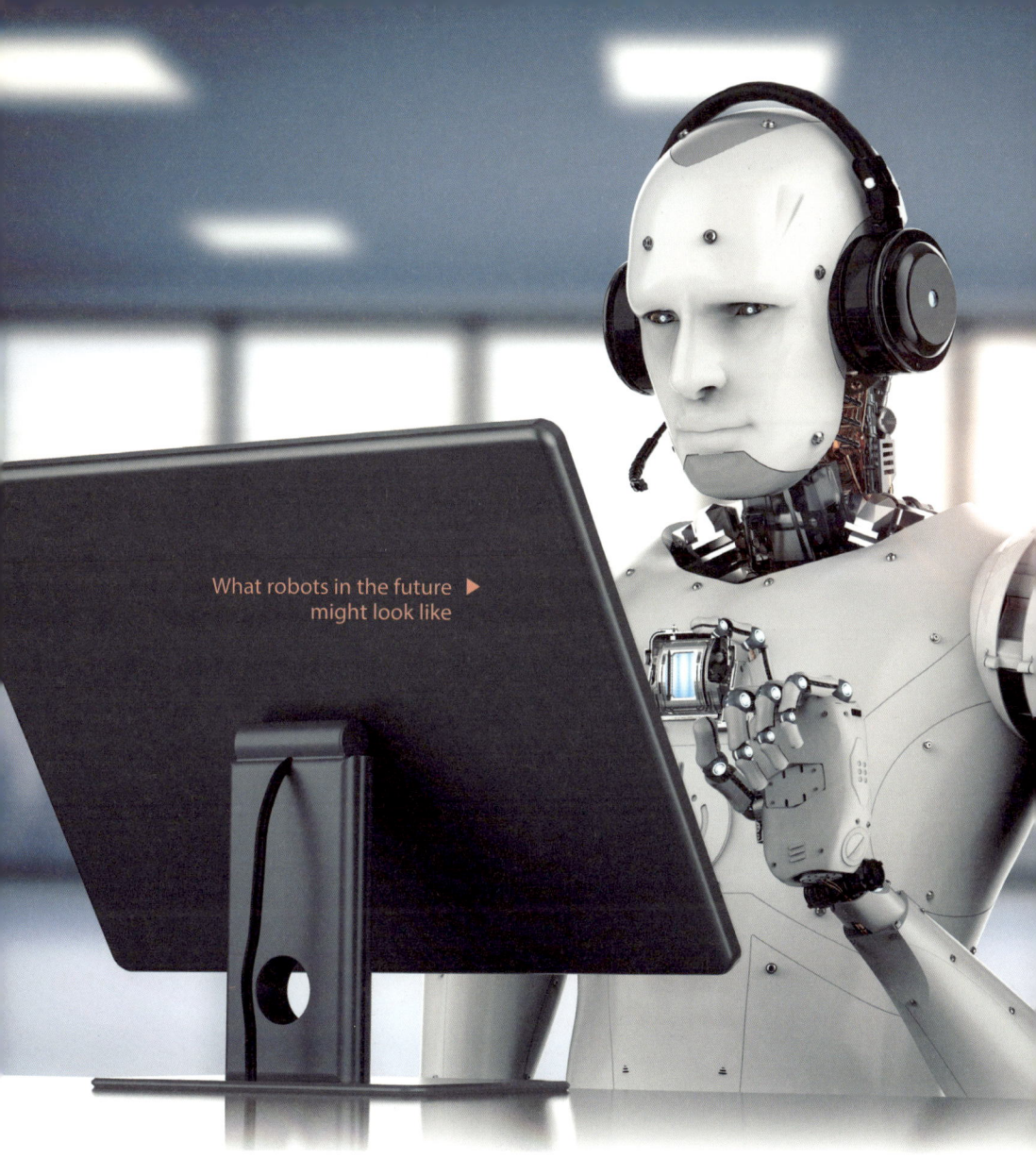

What robots in the future ▶
might look like

In the future, robots might look more like people.
Then anyone will feel fine working with them.

◀ A robot that explores under water

Robots not only work.
They also explore.
Some robots explore dangerous
places where people can't go.
They explore in space or
under the water.

A robot that ▶
explores on Mars

They help scientists learn about these places.

Then people may be able to go there in the future!

A robot exploring a ▶
ship on the bottom
of the sea

Teaching students about robots

Some robotics engineers work in universities.
These engineers teach students about robots.
A few universities now have degree programs called
robotics engineering.
More and more universities will have these kinds of
programs in the future.

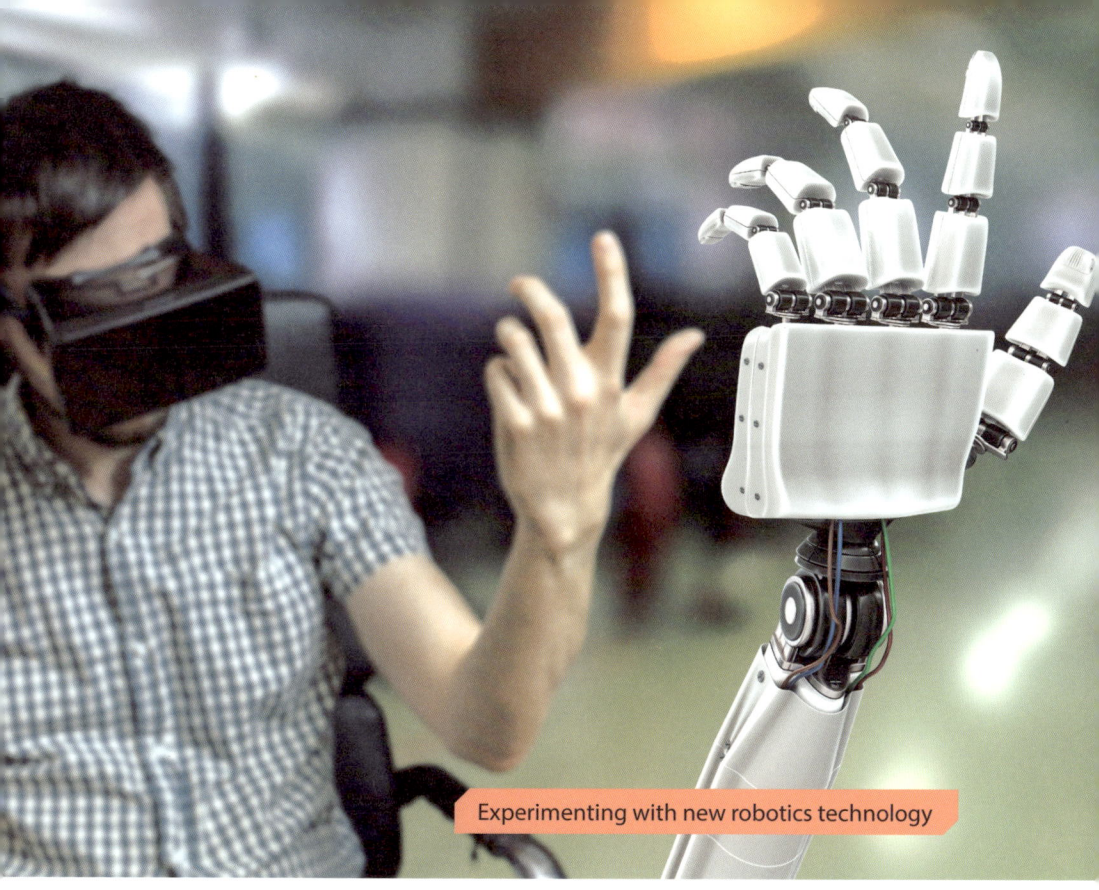

Experimenting with new robotics technology

Robotics engineers in universities do not just teach. They also study new ways to make and use robots. These engineering teachers do experiments to learn more about robots.

Students often help with these experiments. This is a good way for students to learn about robots.

Looking to the Future

Robots are being used in more places every year. Stores and hotels are using them. Museums and parks are using them. Farms are using them.

More and more robotics engineers are needed to make the kinds of robots these places want.

A robot that provides information at a robotics conference ▶

A future robotics engineer

Does this job look interesting to you?

Maybe you can be a robotics engineer in the future!

Comprehension Questions

1. Robotics engineers need to be . . .
   (a) clean and kind.
   (b) creative and patient.
   (c) simple and nice.
   (d) small and careful.

2. Where can young people learn about robots before they go to university?
   (a) In contests
   (b) In robotics summer camps
   (c) In special programs
   (d) All of the above

3. Who do assistant engineers learn from at companies?
   (a) Experienced engineers
   (b) Only people with Ph.D.s
   (c) Master degree students
   (d) All of the above

4. Today, most robots can be found in factories that make . . .
   (a) cars.
   (b) computers.
   (c) things for doctors.
   (d) robots.

5. What is true about robotics engineering programs today?
   (a) The experiments done in them are dangerous.
   (b) Not many universities have them.
   (c) They are only for Ph.D. students.
   (d) All of the above

Key 1. (b) 2. (d) 3. (a) 4. (a) 5. (b)

Glossary

- **assistant** (n.) a person who helps another to do a job

- **contest** (n.) a race or event with a prize for the best

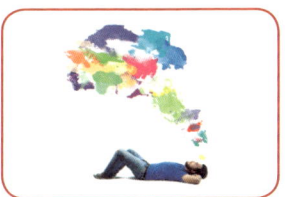

- **creative** (adj.) having the ability to make or think of new things

- **degree** (n.) a title given to a student at the end of a course of study

- **engineer** (n.) a person who thinks of new ways to build things or solve problems

- **experiment** (n.) an organized plan for testing an idea, usually written down to be shared with others

- **explore** (v.) to go into a new place to find out about it

- **factory** (n.) a building in which a product is made ready for sale

- **program** (n.) a plan or schedule of classes that leads to a certificate or degree

- **university** (n.) a place where students study after high school for an advanced degree

Notes

Here are some people currently working as robotics engineers. Readers may enjoy researching these people to learn more about those who are already working in this field.

Brian Gerkey started a nonprofit organization that makes robotics software open and available for research, education, and product development.

Cynthia Breazeal is a research scientist interested in how to make robots look and act more friendly so that people enjoy interacting with them.

Marc Raibert was a researcher in computer science and artificial intelligence before he started a company that makes running, jumping, climbing, and driving robots.

Colin Angle is the CEO of the company that makes Roomba ®, the world famous house cleaning robot.

Daniela Rus is a robotics researcher who has helped to develop the first robotic fish built using the idea of "soft robotics" (no hard parts), and she continues to work on how to print paper robots.

List of Books

LEVEL 1

1. Robotics Engineers
2. Cyber Security Experts
3. Medical Scientists
4. Social Media Managers
5. Asset Managers

LEVEL 2

1. Drone Pilots
2. App Developers
3. Wearable Technology Creators
4. Computer Intelligence Engineers
5. Digital Modelers

LEVEL 3

1. IoT Marketing Specialists
2. Space Pilots
3. Water Harvesters
4. Genetic Counselors
5. Data Miners

LEVEL 4

1. Database Administrators
2. Nanotechnology Research Scientists
3. Quantum Computer Scientists
4. Agricultural Engineers
5. Intellectual Property Lawyers

"The future of the economy is in STEM. That's where the jobs of tomorrow will be."

James Brown (Executive Director of the STEM Education Coalition in Washington, D.C.)

Data from the US Bureau of Labor Statistics (BLS) support that assertion. Employment in occupations related to STEM—science, technology, engineering, and mathematics—is projected to grow to more than 9 million by 2022 [in the US alone] . . . Overall, STEM occupations are projected to grow faster than the average for all occupations.

from *STEM 101: Intro to Tomorrow's Jobs* US Bureau of Labor Statistics